YOUR KNOWLEDGE HAS VALUE

- We will publish your bachelor's and
  master's thesis, essays and papers

- Your own eBook and book -
  sold worldwide in all relevant shops

- Earn money with each sale

Upload your text at www.GRIN.com
and publish for free

GRIN

Peter Klapper

Adsorption equilibria of di(2-ethylhexyl)phosphoric acid at the water-dodecane-interface. Effects of additional electrolytes

GRIN Verlag

Bibliografische Information der Deutschen Nationalbibliothek:

Die Deutsche Bibliothek verzeichnet diese Publikation in der Deutschen National-
bibliografie; detaillierte bibliografische Daten sind im Internet über http://dnb.d-
nb.de/ abrufbar.

Imprint:

Copyright © 2014 GRIN Verlag GmbH
Druck und Bindung: Books on Demand GmbH, Norderstedt Germany
ISBN: 978-3-656-74267-8

Adsorption equilibria of di(2-ethylhexyl)phosphoric acid at the water-dodecane-interface: Effects of additional electrolytes

Keywords: Di(2-ethylhexyl)phosphoric acid, interfacial activity, adsorption equilibrium, counterion adsorption, pseudo-nonionic modelling, liquid-liquid-interface, interfacial tension

Abstract

Based on the adsorption of both derivatives of di(2-ethylhexyl)phosphoric acid - monomer and anion - a new modelling strategy is presented for the adsorption equilibria and the interfacial tension as the consecutive physical value. The starting point is a description of the adsorption equilibrium of these two interfacial active substances using the Langmuir isotherm. Because of the accumulation of the counterions, which is reproduced by the Stern isotherm, the Gibbs adsorption equation, unlike with the nonionic modelling strategy normally used for this material system, has to be supplemented by the quantity of the importance of counterions. A simple model integrating micelle formation makes it possible to simulate the equilibrium interfacial tension even at high concentrations of interfacially active agents.

1 Introduction

The adsorption of surfactants significantly determines the properties of the interface. In liquid-liquid systems the accumulation of substances at the interface mainly modifies the interfacial tension, the interfacial charge and the interfacial rheology [1,2]. These adsorptive interface modifications have far-reaching consequences on the process design in fluid engineering. For example, phase formations such as the mechanisms of breakage and coalescence, the droplet velocity and the mass transport over the interface are influenced by interfacial adsorption in disperse multiphase systems. If the adsorbed substances are the transition components in the physical extraction or if the interfacially active substances are reactants in the reactive extraction, the description of the adsorption equilibria is very important, because in these cases the processes at the interface and in the interfacial layer guide the extraction process.

The adsorption isotherms characterize the equilibrium composition of the interface by linking the concentrations in the interface with those in the volume phases. These isotherms are fundamental for the formulation of unrestricted adsorption processes, because the enrichment step is performed by the local equilibrium between the interface and the adjacent bulk phase, known as the subsurface. The mass transfer from the separate bulk phases to the subsurface is performed by the kinetics of convection and diffusion and, in aqueous regimes, with electrolytes additional by migration. If the adsorption is delayed by interactions in the interface and in the subsurface, the sorptive mechanism must be described using a kinetic approach instead of the isotherms [3]. This kinetic formulation can be derived from the adsorption isotherms, since the isotherms define the steady state of the sorption kinetics.

The suitability of specific isotherms is verified by adjusting measured profiles of the equilibrium interfacial tension. Their change is related with the interfacial concentration and activity changes in the volume phase by means of the Gibbs adsorption equation [3]. The different modelling strategies are the result of assumptions made about additional interface-active agents as a consequence of previously independent adsorbed components [4].

Despite the industrial use of the cation exchanger di(2-ethylhexyl)phosphoric acid – D2EHPA for short – in metal salt extraction [5] and its known interfacial activity [6], it is still common practice to describe the adsorption as a spontaneous chemical reaction [7,8]. Even if the adsorption takes place without sorptive inhibition, the limited enrichment capacity of the interface is suppressed in these methods. In the subsequent metal salt formation reactions this fact results in a stagnation of the mass transfer as the turbulence intensity in the phase from which is adsorbed is increased. At this phase side transport limitations arise from the limited interfacial concentration.

Even the works which deal with the adsorption of D2EHPA at water-oil interfaces reveal large differences. Miyake et al. [9] and Szymanowski et al. [10] describe the equilibrium interfacial tension during the adsorption of the cation exchanger by the Langmuir isotherm for multicomponent adsorption. They base their description on the same minimal area per adsorbed monomer and anion of the cation exchanger. Even Vandegrift et al. [11] and Shen et al. [12] neglect the influence of anion adsorption at the interface and postulate the adsorption of only one component. None of these authors consider the influence of counterions, as demonstrated by Goanker et al. [6] on the basis of experimental curves of the equilibrium interfacial tension.

2 Experimental

The measurements of the equilibrium interfacial tension at the liquid-liquid interface were carried out at 20°C based on a pendant drop using drop shape analysis. In this method, the profile of a rotationally symmetric drop is recorded as a silhouette and is described by the force balance of interfacial tension, buoyancy and gravity [13]. Since there is a density difference between the droplet and bulk phase, the curvature radii of the drop change due to the adsorption and every drop shape change is definitely assigned an interfacial tension. The equilibrium interfacial tension is obtained as the stationary value of the measured curve of the dynamic interfacial tension.

The measurement setup used is shown in Figure 1: a dosing system consisting of a gas-tight glass syringe whose stroke is controlled by a stepper motor is used to generate the aqueous pendant drop on a Teflon capillary which is inserted into a glass cuvette filled with organic phase. This cuvette is embedded in a temperature-measuring cell.

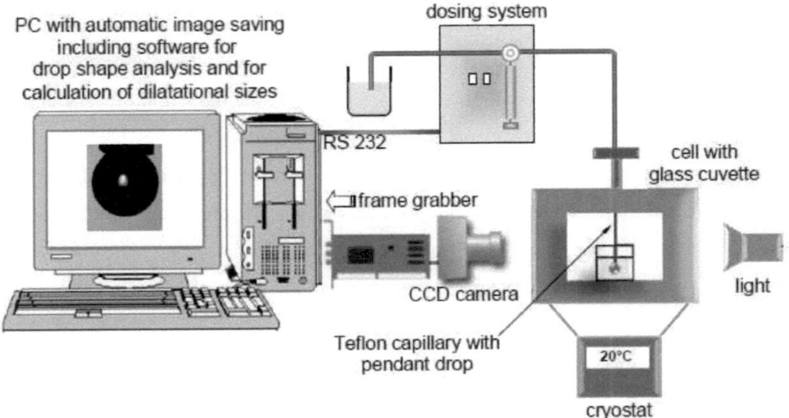

Figure 1: Schematic experimental setup of the pendant-drop tensiometer

The temperature of the cell is stabilized using a cryostat to measure and regulate the temperature in a stirred cooling water (or heating water). The drop is illuminated by a light source whose light is parallelized by a diffuser and lens. The silhouette of the drop is transmitted through the lens with an attached CCD camera to the computer. There it is digitized using a frame grabber [14]. The PAT 1 software, produced by SINTERFACE, allows the digital drop profile to be saved automatically at any time. This software controls the dosing unit via an interface. In this way, the surface or the volume of the drop can optionally either be held constant or modified during the experiment by means of permanent controlling in accordance with the test program. The time profiles of the experimental program are implemented by entering the corresponding mathematical functions.

All chemicals except the used cation exchanger (D2EHPA, 95 weight-%, technical quality, Sigma) were checked for possible contamination using tensiometric measurements. Only in the case of the diluent (n-dodecane, 99 weight-%, for synthesis, Merck) was preparation necessary. This was done by washing with deionised water. All other chemicals were used in analytical quality without any preparation, because no interfacially active impurities were detected.

To study the influence of the two interfacially active variants of the cation exchanger – the monomer and anion – by shifting the dissociation equilibrium and to understand the effect of the counterions on the interfacial tension, sulphuric acid, sodium sulphate or sodium hydroxide were added to the aqueous phase in various concentrations. In the organic dodecane phase the D2EHPA concentration was varied based on the concentration of classical surfactants added at different levels, up to those used in technical applications. Before the interfacial tension was measured of aqueous droplets and an organic volume phase, the two phases were mixed for 24 hours in a shaker with a constant volume ratio. As this completed the mass transfer between the phases, this excluded the possibility of errors resulting from different volumes because of the different shape of drops and the inaccurate filling of the cuvette. So that not all variable concentration values had to be calculated using balance equations, the pH value was additionally measured in the aqueous solution.

3 Modelling

The Gibbs adsorption equation provides the change of interfacial tension in the case of the joint adsorption of the monomer and the cation exchanger anion including the counterion adsorption of protons and sodium ions, if the concentration shift in the electrochemical double layer is neglected:

$$d\gamma = -\Re T \left(\Gamma_{\overline{HR}} \frac{da_{\overline{HR}}}{a_{\overline{HR}}} + \Gamma_{R^-} \frac{da_{R^-}}{a_{R^-}} + \Gamma_{H^+} \frac{da_{H^+}}{a_{H^+}} + \Gamma_{Na^+} \frac{da_{Na^+}}{a_{Na^+}} \right) \tag{1}$$

Because electrostatic effects are left out, the following modelling strategy is called pseudo-nonionic. In Eq. (1) the Langmuir isotherms are introduced with the simplification that the maximum interfacial concentrations of the multicomponent adsorption are the same as in the adsorption of only one component.

$$\Gamma_{\overline{HR}} = \Gamma_{\infty,\overline{HR}} \frac{K_{L,\overline{HR}} a_{\overline{HR}}}{1 + K_{L,\overline{HR}} a_{\overline{HR}} + K_{L,R^-} a_{R^-}} \tag{2}$$

$$\Gamma_{R^-} = \Gamma_{\infty,R^-} \frac{K_{L,R^-} a_{R^-}}{1 + K_{L,\overline{HR}} a_{\overline{HR}} + K_{L,R^-} a_{R^-}} \tag{3}$$

The interfacial concentrations of the counterions are coupled via the Stern isotherm – neglecting the electrostatic interfacial potential – with their bulk phase activities and the interfacial concentration of adsorbed cation exchanger anions.

$$\Gamma_{H^+} = \frac{K_{S,H^+} a_{H^+}}{1 + K_{S,H^+} a_{H^+} + K_{S,Na^+} a_{Na^+}} \Gamma_{R^-} \tag{4}$$

$$\Gamma_{Na^+} = \frac{K_{S,Na^+} a_{Na^+}}{1 + K_{S,H^+} a_{H^+} + K_{S,Na^+} a_{Na^+}} \Gamma_{R^-} \tag{5}$$

Since the monomer concentration of the organic phase is connected with the activities of the proton and the anion of the cation exchanger by the distribution of the monomers in the aqueous phase and the following dissociation, any concentration value of the one component is always defined by the concentration values of the other two components. Therefore, changes to activities of these reactive coupled substances in the Gibbs adsorption equation, Eq. (1), are expressed by the remaining independent variables. The partition equilibrium between monomers in the two phases

$$K_p = \frac{a_{\overline{HR}}}{a_{HR}} \tag{6}$$

and by the mass action law of dissociation

$$K_a = \frac{a_{H^+} \cdot a_{R^-}}{a_{HR}} \tag{7}$$

are used to find the relationship between the different activities:

$$a_{R^-} = \frac{K_a}{K_p} \frac{a_{\overline{HR}}}{a_{H^+}} \tag{8}$$

From this the total differential of the anion activity is formed:

$$da_{R^-} = \frac{K_a}{K_p} \left(\frac{1}{a_{H^+}} da_{\overline{HR}} - \frac{a_{\overline{HR}}}{a_{H^+}^2} da_{H^+} \right) \tag{9}$$

After the isotherms are introduced and the dependent activities are linked, the calculation formula of the equilibrium interfacial tension is obtained by integrating the Gibbs adsorption equation, Eq. (1).

$$\gamma = \gamma_0 - \Re T \frac{\Gamma_{\infty,\overline{HR}} K_{L,\overline{HR}} a_{\overline{HR}} + \Gamma_{\infty,R^-} K_{L,R^-} a_{R^-}}{K_{L,\overline{HR}} a_{\overline{HR}} + K_{L,R^-} a_{R^-}} \ln\left[1 + K_{L,\overline{HR}} a_{\overline{HR}} + K_{L,R^-} a_{R^-} \right]$$

$$- \Re T \frac{\Gamma_{\infty,R^-} K_{L,R^-} a_{R^-} K_{S,H^+} a_{H^+}}{(1 + K_{S,Na^+} a_{Na^+})(1 + K_{L,\overline{HR}} a_{\overline{HR}}) - K_{L,R^-} a_{R^-} K_{S,H^+} a_{H^+}} \ln \frac{1 + \dfrac{K_{L,R^-} a_{R^-}}{1 + K_{L,\overline{HR}} a_{\overline{HR}}}}{1 + \dfrac{K_{S,Na^+} a_{Na^+}}{K_{S,H^+} a_{H^+}}} \tag{10}$$

$$- \Re T \frac{\Gamma_{\infty,R^-} K_{L,R^-} a_{R^-}}{1 + K_{L,\overline{HR}} a_{\overline{HR}} + K_{L,R^-} a_{R^-}} \ln\left[1 + \frac{K_{S,Na^+} a_{Na^+}}{1 + K_{S,H^+} a_{H^+}} \right]$$

Initially this relation applies only to the submicellar concentration range, but the equation can easily be modified to include the micellar range. The modification makes use of the fact that the micelle formation is not a chemical reaction in the classical sense but rather a self-assembly of surfactant molecules. Because of this, the micelle formation has no effects on chemical reaction equilibria processes. Its importance results from its effect on the adsorption isotherms. The disaggregated quantity of the interfacially active component is crucial for the adsorption isotherms. Because of the known micelle formation of the cation exchanger anion [15] and the discussed micellar influences the linkages between the activities are also valid in the micellar concentration range and the anion activity simply has to be replaced in the calculation formula of the equilibrium interfacial tension, Eq. (10), with that of the unbound anion.

$$a_{R^-} = \alpha_{R^-} \frac{K_a}{K_p} \frac{a_{\overline{HR}}}{a_{H^+}} \tag{11}$$

The fraction of the unbound anion is defined by means of a micelle formation reaction, which is technically similar to a chemical reaction. Neglecting the importance of the counterions on the micelle formation, the micelle formation can be described using the reaction scheme:

$$n R^- \Leftrightarrow \left(R^-\right)_n \tag{12}$$

In this reaction scheme, there are additionally assumed to be monodisperse micelles with the same aggregation number. The corresponding mass action law leads to:

$$K_{m,R^-} = \frac{a_{\left(R^-\right)_n}}{a_{R^-}^n} \tag{13}$$

Since the micelle formation does not affect the chemical reaction equilibria, the micelle activity is formulated from the equilibria of the partition and the dissociation using the fraction of the unbound anion.

$$a_{\left(R^-\right)_n} = (1 - \alpha_{R^-}) \frac{K_a}{K_p} \frac{a_{\overline{HR}}}{a_{H^+}} \tag{14}$$

By introducing this equation, Eq. (14), together with the relation in Eq. (11), into the mass action law, Eq. (13), the fraction of the unbound anion is defined by an implicit equation:

$$\frac{1 - \alpha_{R^-}}{\alpha_{R^-}^n} = K_{m,R^-} \left(\frac{K_a}{K_p} \frac{a_{\overline{HR}}}{a_{H^+}} \right)^{n-1} \tag{15}$$

While the activity of the proton was measured, the other activities have to be calculated. Since sodium is not extracted by D2EHPA to any noticeable degree, the concentration of the sodium ion is known. The other, unknown concentration values have to be determined using mass balances and reaction equilibria. The unknown activity coefficient is described involving the radius of the hydrated ion as the only ion-specific parameters according to the extended Debye-Hückel law. For the activity coefficients of an arbitrary z-valent ion, this produces [16]:

$$\ln \gamma_i = -\frac{z_i^2 \Im^2 \cdot \kappa}{2\varepsilon_0 \varepsilon_{rel} \Re T N_A \left(1 + \kappa \cdot r_{i,hyd}\right)} \tag{16}$$

The Debye-Hückel length this includes is defined by means of the ionic strength.

$$\kappa = \sqrt{\frac{8\pi\Im^2}{\varepsilon_0 \varepsilon_{rel} \Re T}} \cdot \sqrt{I} \qquad \text{with} \qquad I = \tfrac{1}{2}\sum_i z_i^2 \cdot c_i \tag{17}$$

Because of the dependence of the electrolyte activity coefficients on the ionic strength, all the ion concentrations have to be determined. Since these values are derived from the different reaction-specific equations of mass action law, knowledge is required of the activity coefficients of all the ions present. The hydrated ion radii (see Tab. 1) are calculated from the average electrolyte activities [17] using a numerical regression procedure.

H^+	Na^+	OH^-	HSO_4^-	SO_4^{2-}	R^-
4.3 Å	3.5 Å	4.8 Å	3.9 Å	3.7 Å	3.9 Å

Table 1: Used hydrated ionic radii r_{hyd}

The radius of the hydrated cation exchanger anion is estimated as the average of other values. The determination of the concentration values, which are necessary for the calculation of the ionic strength, requires only the presence of the dissociated forms of the sulphuric acid. The equilibrium between the sulphate and hydrogen sulphate ions is formulated via the mass action law

$$K_{HSO_4^-} = \frac{c_{SO_4^{2-}} \cdot a_{H^+}}{c_{HSO_4^-}} \cdot \frac{\gamma_{SO_4^{2-}}}{\gamma_{HSO_4^-}} \tag{18}$$

and it is calculated by assessing the total amount of sulphate. The concentration of hydroxide, which is also unknown, is defined thus:

$$c_{OH^-} = \frac{K_{H_2O}}{\gamma_{OH^-} \cdot a_{H^+}} \tag{19}$$

The required dissociation constants are taken from the standard references [18,19] describing the equilibria of inorganic electrolytes.

The activity of the cation exchanger anion is explained by Eq. (8). From this the corresponding concentration is calculated by linking:

$$c_{R^-} = \frac{a_{R^-}}{\gamma_{R^-}} \tag{20}$$

At first the activity of the monomer has to be determined. Taking account of the equilibrium of the dimerization

$$K_d = \frac{a_{\overline{(HR)_2}}}{a_{\overline{HR}}^2} \tag{21}$$

this activity is expressed on the basis of the total D2EHPA concentration and the measured proton activity using the mass balance, which includes all chemical D2EHPA variations in the organic phase and in the aqueous phase.

$$a_{\overline{HR}} = \frac{\gamma_{\overline{(HR)_2}}}{4K_d \gamma_{\overline{HR}}^\infty} \left(\sqrt{ \left(1 + \frac{V_w}{V_o} \frac{\gamma_{\overline{HR}}^\infty}{K_p} \left[\frac{K_a}{\gamma_{R^-} a_{H^+}} + \frac{1}{\gamma_{\overline{HR}}^\infty} \right] \right)^2 + \frac{8 K_d \gamma_{\overline{HR}}^{\infty\,2}}{\gamma_{\overline{(HR)_2}}} c_{D2EHPA,0} } \right.$$
$$\left. - \left(1 + \frac{V_w}{V_o} \frac{\gamma_{\overline{HR}}^\infty}{K_p} \left[\frac{K_a}{\gamma_{R^-} a_{H^+}} + \frac{1}{\gamma_{\overline{HR}}^\infty} \right] \right) \right) \tag{22}$$

For this equation the total D2EHPA concentration is noted down as a function of the monomer concentration.

According to current knowledge [20] the quantity of the monomer is negligible in aliphatic diluents in comparison with the quantity of the dimer. Hence the concentration dependence of the dimer activity coefficient can be determined by means of the Wilson model [21] for the binary mixture of dodecane-dimer.

$$\ln \gamma_i = 1 - \ln \left(\sum_j x_j \cdot \Lambda_{ij} \right) - \sum_k \frac{x_k \cdot \Lambda_{ki}}{\sum_j x_j \cdot \Lambda_{kj}} \quad \text{with} \quad x_i = \frac{c_i}{\sum_j c_j} \tag{23}$$

The calculated interaction parameters of the Wilson model [14] are listed in Table 2.

	$C_{12}H_{26}$	$(HR)_2$
$C_{12}H_{26}$	1	3.2173
$(HR)_2$	0.0381	1

Table 2: Interaction parameter Λ of the Wilson model

Because the constants of the adsorption isotherms and the chemical equilibrium constants were obtained from data fitting the measured equilibrium interfacial tension, though the values of the constants can be indicated (see Tab. 3). As a result of the combined fitting all the parameters depend on the model of the adsorption equilibria. In principle the chemical equilibrium constants are, however, absolutely independent of the choice of the adsorption equilibrium model. This situation does not change the fact that, with the proposed modelling strategy, better data fitting is achieved than with the conventional modelling. Perhaps for certain constants of the chemical equilibria a better description of the equilibrium interfacial tension could be achieved using the ionic modelling strategy.

$\Gamma_{\infty,\overline{HR}}$	$K_{L,\overline{HR}}\cdot\gamma^{\infty}_{\overline{HR}}$	Γ_{∞,R^-}	K_{L,R^-}	K_{m,R^-}	n
$2.596\cdot10^{-6}$	$3.620\cdot10^{5}$	$1.632\cdot10^{-6}$	$1.614\cdot10^{6}$	$4.956\cdot10^{30}$	14
$\mathrm{mol/m^2}$	$\mathrm{l/mol}$	$\mathrm{mol/m^2}$	$\mathrm{l/mol}$	$(\mathrm{mol/l})^{1-n}$	
$K_d\dfrac{\gamma^{\infty\,2}_{\overline{HR}}}{\gamma^{\infty}_{\overline{(HR)_2}}}$	$K_a\cdot\gamma^{\infty}_{HR}$	$K_p\dfrac{\gamma^{\infty}_{HR}}{\gamma^{\infty}_{\overline{HR}}}$	K_{S,H^+}	K_{S,Na^+}	
$1.231\cdot10^{2}$	$1.231\cdot10^{2}$	$2.557\cdot10^{6}$	$1.428\cdot10^{2}$	$6.913\cdot10^{2}$	
$\mathrm{l/mol}$	$\mathrm{mol/l}$		$\mathrm{l/mol}$	$\mathrm{l/mol}$	

Table 3: Fitted parameters of the multicomponent adsorption and the chemical equilibria

But in the case of the additional fitting of the chemical constants the ionic strategy does not offer better data fitting [14]. For the choice of the basic modelling strategy the type of isotherms and the description of the micelle formation are more important.

4 Results

The data fitting (lines) provides a good description of the measurement on the basis of the discussed model. The statistical error sizes are very small, for example. Despite the very different experimental conditions, the standard deviation is only 7.6 % with a mean relative error of 6.2 %. The good theoretical approximation of all the measured curves is confirmed by the maximum relative error of 23.5 %. If the micelle formation is neglected in the model, the standard deviation increases sixfold, since the two measured series with the high concentrations of added sodium hydroxide dominate the fitting results.

Under the condition of added sulphuric acid (see Fig. 2) the comparison of the measured and the calculated interfacial tension curves demonstrates that the low concentration sensitivity is reproduced very well. The remaining differences should primarily be a result of using the Langmuir isotherm in the low concentration range of the cation exchanger.

Even with the addition of sodium sulphate the calculated and measured curves match almost perfectly for total D2EHPA concentrations greater than 1 µmol/l (see Fig. 3). Only the test series with the addition of 1 mmol/l sodium sulphate have larger deviations. In this case too, it can be seen that the model can both describe a trend and provide a quantitative description.

The curves with additional sodium hydroxide (see Fig. 4) demonstrate that the model can provide a satisfactory reproduction of the interfacial tension with increasing basicity. This result is noteworthy because of the simplified assumptions regarding micelle formation, considering only the dominant form of aggregation. Although the differences between the measurement and simulation are greater here, the trends are correct for all concentration values and the curves are quantitatively compliant for the two extreme concentrations of sodium hydroxide.

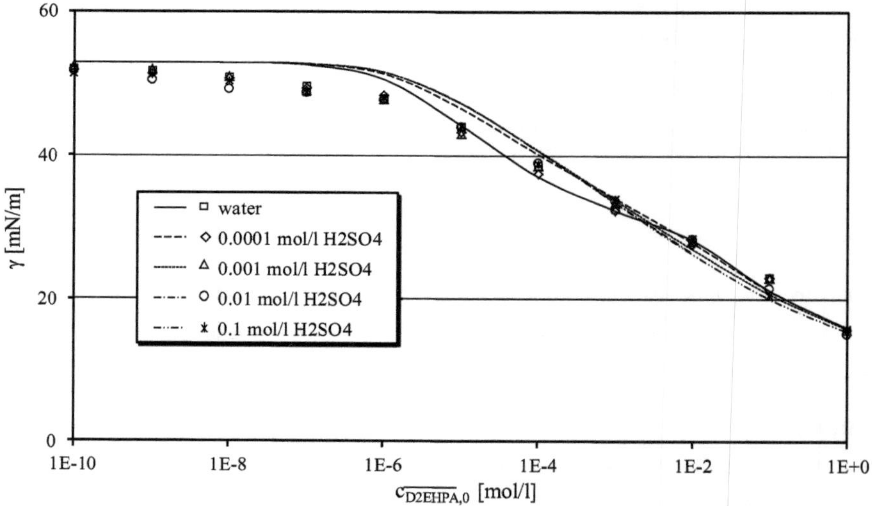

Figure 2: Equilibrium interfacial tension for various added sulphuric acid concentrations

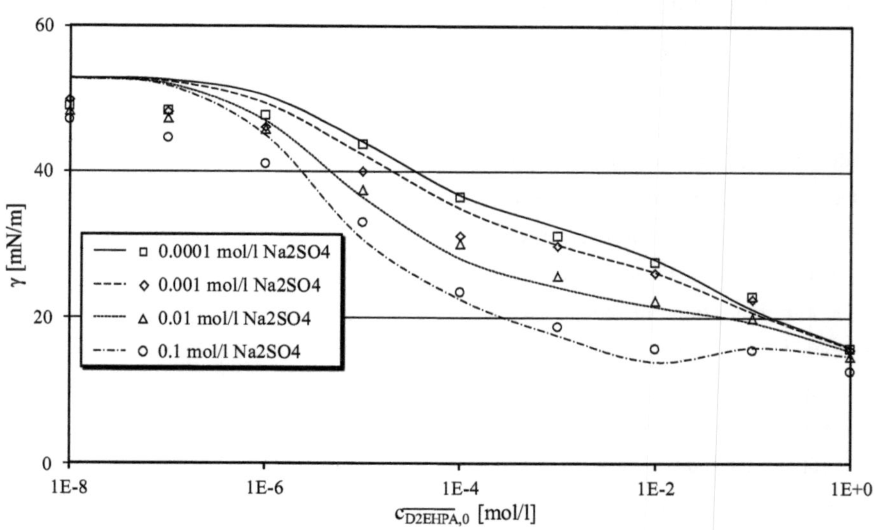

Figure 3: Equilibrium interfacial tension for various added sodium sulphate concentrations

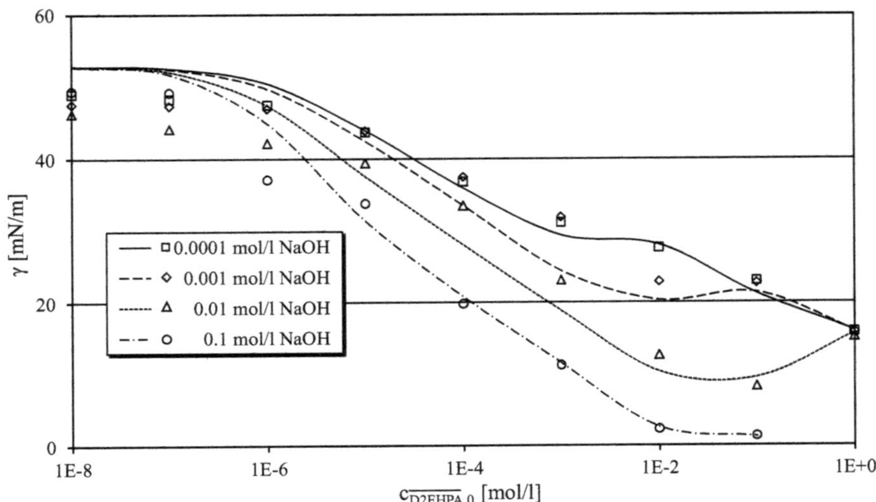

Figure 4: Equilibrium interfacial tension for various added sodium hydroxide concentrations

In spite of the data fitting the measured equilibrium interfacial tensions illustrate the different interfacial activity of the cation exchanger anion and the monomer. Furthermore, the specific influences of the counterion adsorption are clarified. These factors induce locally extreme interfacial tension values if sodium sulphate or sodium hydroxide is added.

Because of the dissociation equilibrium of the cation exchanger the measured values are primarily caused by the adsorption of the monomer in the cases with and without the addition of sulphuric acid (see Fig. 2). Therefore, the interfacial tension curves are similar and they decrease as the total D2EHPA concentration increases.

The addition of sodium sulphate buffers the pH because of the bisulphate-sulphate balance. Thus, as a result of cation exchanger dissociation, more anions from the organic phosphoric acid are formed in the aqueous phase than in sulphuric acid solutions. This leads to an enhanced decrease in the interfacial tension with the sodium sulphate concentration is increased (see Fig. 3). As the cation exchanger concentration is increased, the buffer effect is exhausted even when high sodium sulphate concentrations are added, and the anion adsorption becomes less important in comparison with the monomer adsorption. However, even at high concentrations of D2EHPA, more D2EHPA is always dissociated than without added sodium sulphate. Therefore, only at the two low sodium sulphate concentrations are the interfacial tensions nearly the same as in the sulphate system, if the total D2EHPA concentration is greater than 0.1 mol/l. In the case of the 0.1 molar sodium sulphate solution the interfacial tension curve has a saddle point at the 0.1

molar concentration of D2EHPA because of the changes in the interfacial concentrations. The special importance of the sodium concentration on the interfacial tension curves can especially be seen from the interfacial tension reductions for D2EHPA concentrations of less than 1 mmol/l, because the concentration changes caused by the acid-base balance of the sulphate-bisulphate equilibrium are insignificant. The interfacial tensions are severely reduced in the solution with the higher sodium sulphate concentration. This effect is caused among other things by the counterion adsorption, which follows the adsorption of the cation exchanger anion.

As a result of the acid-base reaction between the cation exchanger and the sodium hydroxide the cation exchanger anion is transferred to the aqueous phase and is then adsorbed increasingly at the interface. At the same time the cation exchanger and the monomeric type of D2EHPA wear down in the organic phase. As long as the sodium hydroxide concentration is sufficiently large, the interfacial tension change is determined by the adsorption of the anion and the associated sodium ion. The pH value decreases if the initial D2EHPA concentration is increased. This leads to an increase in the undissociated fraction of the cation exchanger (monomer) at the interface and in the organic phase. Therefore, the interfacial tension rises. Further increases in the D2EHPA concentration cause increased monomer adsorption at the interface and therefore the interfacial tension diminishes. For these reasons, in the case of the addition of sodium hydroxide (see Fig. 4) for an 0.001 molar sodium hydroxide solution, the interfacial tension curves show both a weak local minimum and a maximum. For the 0.01 molar sodium hydroxide solution a distinct minimum is found. If a 0.1 molar sodium hydroxide solution is brought into contact with the one molar D2EHPA solution, a stable microemulsion is formed. That is why the interfacial tension cannot be measured under this experimental condition.

5 Summary

The equilibrium interfacial tension of the joint adsorption of the monomer and anion in the cation exchanger D2EHPA is successfully simulated using a pseudo-nonionic model including the electrolyte influence. The adsorption of the two individually interfacially active substances is described using the Langmuir isotherm of multicomponent adsorption. The involvement of the counterion adsorption during the adsorption of the cation exchanger anion is modelled using the Stern isotherm. If the chemical equilibrium is not affected by micelle formation, the micelle formation of the cation exchanger anion can be formulated in a monodisperse manner. Any possible influence of the interfacial potential on the adsorption equilibria and thus on the interfacial tension is neglected.

Both the low sensitivity regarding the sulphuric acid concentration and the strong influence of sodium ions in the case of added sodium sulphate can be reflected correctly by the presented model. Moreover, it is possible to describe well the measured interfacial tensions by involving a simple model of micelle formation even in the case of sodium hydroxide addition.

Both the measured and the calculated curves of the equilibrium interfacial tension confirm that there is strong interfacial activity of the anion in the cation exchanger D2EHPA. Far more important, however, are the facts that the adsorption of the counterions seriously affects the interfacial tension and that many of the effects attributed to the anion of the organic phosphoric acid are obviously caused by the adsorption of the counterions.

Nomenclature

Symbols

a	activity [mol/l]
c	concentration [mol/l]
I	ionic strength [mol/l]
K	adsorption and chemical equilibrium constant [depends]
n	stoichiometric factor of micelle formation [-]
r	radius [m]
T	temperature [K]
V	volume [m^3]
x	mol fraction [-]
z	ionic charge [-]
γ	interfacial tension, activity coefficient [depends]
Γ	interfacial concentration [mol/m^2]
Δ	difference operator
κ	reciprocal Debye-Hückel length [1/m]
Λ	Wilson's interaction parameter [-]

Subscripts and superscripts

a	dissociation
d	dimerization
hyd	hydrated
i, j, k	species
L	Langmuir
m	micelle
o	organic phase
p	partition equilibrium
S	Stern

w		aqueous phase
0		starting size
∞		infinite dilution, maximum

Constants

N_A		Avogadro's number ($6.0225 \cdot 10^{23}$ 1/mol)
ε_0		dielectric constant of vacuum ($8.8544 \cdot 10^{-12}$ C^2/Nm^2)
T		Faraday constant ($9.6487 \cdot 10^4$ C/mol)
Y		ideal gas constant (8.3143 J/(K$\cdot$mol))

Appendix

After the introduction of the Stern isotherms, Eq. (4) and Eq. (5), and the activity relations, Eq. (8) and Eq. (9), the Gibbs adsorption equation Eq. (1) can be formulated thus:

$$d\gamma = -\Re T \left[\left(\Gamma_{\overline{HR}} + \Gamma_{R^-} \right) \frac{da_{\overline{HR}}}{a_{\overline{HR}}} - \frac{1 + K_{S,Na^+} a_{Na^+}}{1 + K_{S,H^+} a_{H^+} + K_{S,Na^+} a_{Na^+}} \Gamma_{R^-} \frac{da_{H^+}}{a_{H^+}} \right.$$
$$\left. + \frac{K_{S,Na^+}}{1 + K_{S,H^+} a_{H^+} + K_{S,Na^+} a_{Na^+}} \Gamma_{R^-} da_{Na^+} \right] \tag{A1}$$

As soon as the two interfacial activities are replaced by the Langmuir isotherms, Eq. (2) and Eq. (3), the resultant differential equation is solved by partial integration.

Integrating the first integral, which describes the direct influence of the adsorbed monomers and the cation exchanger anions, leads to:

$$\Delta\gamma_{\overline{HR},R^-} = -\Re T \int \left(\Gamma_{\overline{HR}} + \Gamma_{R^-} \right) \frac{da_{\overline{HR}}}{a_{\overline{HR}}}$$
$$= -\Re T \frac{\Gamma_{\infty,\overline{HR}} K_{L,\overline{HR}} a_{\overline{HR}} + \Gamma_{\infty,R^-} K_{L,R^-} a_{R^-}}{K_{L,\overline{HR}} a_{\overline{HR}} + K_{L,R^-} a_{R^-}} \ln\left(1 + K_{L,\overline{HR}} a_{\overline{HR}} + K_{L,R^-} a_{R^-} \right) \tag{A2}$$

The influence of the proton adsorption is described using the following integral:

$$\Delta\gamma_{H^+} = \Re T \int \frac{1 + K_{S,Na^+} a_{Na^+}}{1 + K_{S,H^+} a_{H^+} + K_{S,Na^+} a_{Na^+}} \Gamma_{R^-} \frac{da_{H^+}}{a_{H^+}} + c_1$$
$$= \Re T \int \frac{\Gamma_{\infty,R^-} K_{L,R^-} \frac{K_a}{K_p} a_{\overline{HR}}}{a_{H^+}^2 (1 + K_{L,\overline{HR}} a_{\overline{HR}}) + K_{L,R^-} \frac{K_a}{K_p} a_{\overline{HR}} a_{H^+}} \cdot da_{H^+} \tag{A3}$$
$$- \Re T \int \frac{\Gamma_{\infty,R^-} K_{L,R^-} \frac{K_a}{K_p} a_{\overline{HR}}}{a_{H^+} (1 + K_{L,\overline{HR}} a_{\overline{HR}}) + K_{L,R^-} \frac{K_a}{K_p} a_{\overline{HR}}} \cdot \frac{K_{S,H^+}}{1 + K_{S,H^+} a_{H^+} + K_{S,Na^+} a_{Na^+}} da_{H^+} + c_1$$

To integrate the first integral the following substitutions must be made:

$$A^* = 1 + K_{L,\overline{HR}}\, a_{\overline{HR}} \qquad (A4a)$$

$$B^* = K_{L,R^-}\, \frac{K_a}{K_p}\, a_{\overline{HR}} \qquad (A4b)$$

Substitutions are also necessary for the integration of the second integral:

$$A = K_{S,H^+}\,(1 + K_{L,\overline{HR}}\, a_{\overline{HR}}) \qquad (A5a)$$

$$B = (1 + K_{S,Na^+}\, a_{Na^+})\,(1 + K_{L,\overline{HR}}\, a_{\overline{HR}}) + K_{L,R^-}\, \frac{K_a}{K_p}\, a_{\overline{HR}}\, K_{S,H^+} \qquad (A5b)$$

$$C = K_{L,R^-}\, \frac{K_a}{K_p}\, a_{\overline{HR}}\,(1 + K_{S,Na^+}\, a_{Na^+}) \qquad (A5c)$$

In this way, the integrated functions are derived by the tabulated indefinite integrals [22] and the proton influence can be defined by:

$$\Delta\gamma_{H^+} = \Re T \Gamma_{\infty,R^-} \left(\int \frac{B^*}{a_{H^+}(A^*\, a_{H^+} + B^*)}\, da_{H^+} - \int \frac{B^* K_{H^+}}{A\, a_{H^+}^2 + B\, a_{H^+} + C}\, da_{H^+} \right) + c_1$$

$$= -\Re T \Gamma_{\infty,R^-} \left(\ln \frac{A^*\, a_{H^+} + B^*}{a_{H^+}} + \frac{B^* K_{H^+}}{\sqrt{B^2 - 4AC}}\, \ln \frac{2A\, a_{H^+} + B - \sqrt{B^2 - 4AC}}{2A\, a_{H^+} + B + \sqrt{B^2 - 4AC}} \right) + c_1 \qquad (A6)$$

The determination of the sodium influence on the interfacial tension change is simple:

$$\Delta\gamma_{Na^+} = -\Re T \int \frac{K_{S,Na^+}}{1 + K_{S,H^+}\, a_{H^+} + K_{S,Na^+}\, a_{Na^+}}\, \Gamma_{R^-}\, da_{Na^+} + c_2$$

$$= -\Re T \Gamma_{R^-}\, \ln\!\left(1 + K_{S,H^+}\, a_{H^+} + K_{S,Na^+}\, a_{Na^+}\right) + c_2 \qquad (A7)$$

The integration constants are calculated by determining the limit value with the zero convergence of the associated activities. In this case, the interfacial tension changes caused by the counterion adsorption must vanish. An additional criterion must be met: if the activity of the cation exchanger anion approaches zero, the interfacial tension changes are defined only by the non-ionic part. Summation of the three interfacial tension parts yields the relationship in Eq. (10).

Based on the initial D2EHPA concentration in the monomer form, for different volumes of the oil and water phase the D2EHPA mass balance is:

$$V_o \cdot c_{\overline{HDEHP},0} = V_o \cdot c_{\overline{HDEHP}} + V_w \cdot (c_{R^-} + c_{HR}) \qquad (A8)$$

After the aqueous monomer activity is replaced by a linear concentration-dependent relation, the fraction of the undissociated organic phosphoric acid which is solved in the aqueous solution is described using the relation in Eq. (6) as a function of the monomer activity in the organic solution.

$$c_{HR} = \frac{a_{\overline{HR}}}{K_p \cdot \gamma_{\overline{HR}}^{\infty}}$$ (A9)

In sum, the total concentration of D2EHPA in the organic phase is composed of the parts of the monomer and the dimer. Using the equilibrium relationship of the dimerization, Eq. (21), and the concentration independence of the monomer activity coefficient, the following equation is obtained:

$$c_{\overline{D2EHPA}} = \frac{a_{\overline{HR}}}{\gamma_{\overline{HR}}^{\infty}} + \frac{2K_d \cdot a_{\overline{HR}}^2}{\gamma_{\overline{(HR)_2}}}$$ (A10)

Introducing this equation into Eq. (A8) together with the dissociation relation, Eq. (8), in combination with the definition of the anion concentration, Eq. (20), and the relation of the undissociated D2EHPA, Eq. (A9) the D2EHPA, the initial concentration can be defined thus:

$$c_{\overline{D2EHPA},0} = \frac{2K_d}{\gamma_{\overline{(HR)_2}}} \cdot a_{\overline{HR}}^2 + \left(\frac{1}{\gamma_{\overline{HR}}^{\infty}} + \frac{V_w}{V_o} \left[\frac{K_a}{K_p} \frac{1}{\gamma_{R^-} a_{H^+}} + \frac{1}{K_p \gamma_{\overline{HR}}^{\infty}} \right] \right) \cdot a_{\overline{HR}}$$ (A11)

Regarding the monomer activity, this relation is a quadratic equation which, when solved by completing the squares, leads to the desired correlation in Eq. (22).

Literature

[1]   Hunter, R. J.: *Foundations of colloid science*; Clarendon Press, 2001

[2]   Miller, R.; Wüstneck, R.; Krägel, J.; Kretzschar, G.: *Dilational and shear rheology of adsorption layers at liquid interfaces*; Colloids Surfaces A 111 (1996) 75-118

[3]   Dukhin, S. S.; Kretzschmar, G.; Miller, R.: *Dynamics of adsorption at liquid interfaces*; Studies Interface Sci. Vol. 1, Elsevier, 1995

[4]   Prosser, A. J.; Franses, E. I.: *Adsorption and surface tension of ionic surfactants at the air-water interface: Review and evaluation of equilibrium models*; Colloids Surfaces A 178 (2001) 1-40

[5]   Lo, T. C. (Hrsg.); Baird, M. H. I. (Hrsg.); Hanson, C. (Hrsg.): *Handbook of solvent extraction*; Krieger Publishing Company, 1991

[6]   Gaonkar, A. G.; Neuman, R. D.: *Interfacial activity, extractant selectivity and reversed micellization in hydrometallurgical liquid/liquid extraction systems*; J. Colloid Interface Sci. 119 (1987) 251-261

[7]   Ji, J.; Mensforth, K. H.; Perera, J. M.; Stevens, G. W.: *The role of kinetics in the extraction of zinc with D2EHPA in a packed column*; Hydrometallurgy 84 (2006) 139-148

[8]   Mansur, M. B.; Slater, M. J.; Biscaia, E. C.: *Kinetic analysis of the reactive liquid-liquid test system ZnSO₄/D2EHPA/n-heptane*; Hydrometallurgy 63 (2002) 107-116

[9]   Miyake, Y.; Harada, M.: *Extraction rate of metal ions with acidic organophosphorus extractant*; Rew. Inorg. Chem. 10 (1989) 65-92

[10]   Szymanowski, J.; Cote, G.; Blondet, I.; Bouvier, C.; Bauer, D.; Sabot, J. L.: *Interfacial activity of bis(2-ethylhexyl)phosphoric acid in model liquid-liquid extraction systems*; Hydrometallurgy 44 (1997) 163-178

[11]   Vandegrift, G. F.; Horwitz, E. P.: *Interfacial activity of liquid-liquid extraction reagents - I*; J. Inorg. Nucl. Chem. 42 (1980) 119-125

[12]   Shen, J.; Gao, Z.; Xi, Z.; Sun, S.: *The interfacial properties and metal extraction kinetics of some organophosphorus extractants*; Proc. ISEC′86, München, Vol. II 287-293

[13]   Möbius, D. (Hrsg.); Miller, R. (Hrsg.): *Drops and bubbles in interfacial research*; Studies Interface Sci. Vol. 6, Elsevier, 1998

[14]   Klapper, P.: *Tensiometrische Stofftransportuntersuchungen der Zinkextraktion mit dem Kationenaustauscher Di(2-ethylhexyl)phosphorsäure*; Thesis, TU Bergakademie Freiberg, 2010

[15]   Shioi, A.; Harada, M.; Tanabe, M.: *X-ray and light scattering from oil-rich microemulsions containing sodium bis(2-ethylhexyl) phosphate*; Langmuir 12 (1996) 3201-3205

[16]   Pitzer, K. S.: *Thermodynamics*; McGraw-Hill, 1995

[17]   D′Ans, J.; Lax, E.: *Taschenbuch für Chemiker und Physiker*; Springer, 1943

[18]   Högfeldt, E.: *Stability constants of metal-ion complexes. Part A: Inorganic ligands*; Pergamon Press, 1982

[19]   Sillen, L. G.; Martell, A. E.: *Stability constants of metal-ion complexes*; The chemical society, 1964

[20]   MacLean, D. W. J.; Dreisinger, D. B.: *The kinetics of zinc extraction in the di(2-ethylhexyl)phosphoric acid, n-heptane-Zn(ClO₄)₂, HClO₄, H₂O system using the rotating diffusion cell*; Hydrometallurgy 33 (1993) 107-136

[21]   Prausnitz, J. M.; Lichtenthaler, R. N.; de Azevedo, E. G.: *Molecular thermodynamics of fluid-phase equilibria*; Prentice Hall, 1999

[22]   Bronstein, I. N.; Semendjajew, K. A.: *Handbook of mathematics*; Springer, 2004